From Kiki..

Steamboats on the River Coloring Book

Written and illustrated by Joseph A. Arrigo

Merry Christmas!!

2019

PELICAN PUBLISHING COMPANY
Gretna 2012

To the Delta Queen's *Mimi Mills, Sheila Michel, and Marsha Opsata-Sparks for their help and support*

Copyright © 1997
By Joseph A. Arrigo
All rights reserved

First printing, November 1997
Second printing, November 2003
Third printing, December 2012

The word "Pelican" and the depiction of a pelican are trademarks of Pelican Publishing Company, Inc., and are registered in the U.S. Patent and Trademark Office.

ISBN: 9781565543164

Printed in the United States of America
Published by Pelican Publishing Company, Inc.
1000 Burmaster Street, Gretna, Louisiana 70053

The large steamboat *J. M. White* steams past the New Orleans riverfront at the French Quarter.

Passengers boarded the steamboat *J. M. White* from the main deck, which was also used to carry cargo.

Two small sternwheelers along the New Orleans riverfront.

One of nine boats named *Natchez,* with its stage (gangplank) lowered and ready for bales of cotton to be loaded.

The newest steamboat named *Natchez,* which operates daily, mostly taking tourist passengers along the New Orleans riverfront.

Passengers enjoyed a steamboat race. Racing between steamboats was, and still is, popular.

The main salon of most steamboats was like the lobby of a fancy hotel.

Early steamboats were actually steered by turning the ship's wheel in the pilot house.

In newer steamboats, the steering is done by controls and the wheel is mainly for show.

The *U.S. Snag* was a steamboat designed to remove tree and log snags from the river.

The steam powered towboat *J. B. O'Brien* worked in the Natchez, Mississippi area for more than 20 years.

Before bridges were built across the river, steamboat ferries carried trains from one side to the other.

The paddle wheel of the historic steamboat *Delta Queen*.

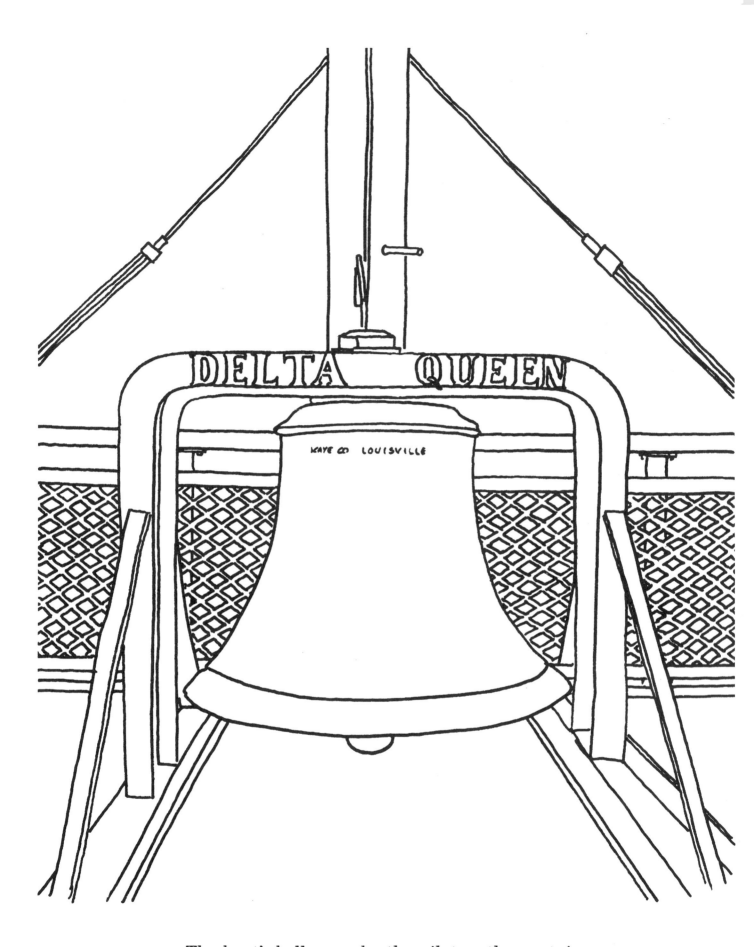

The boat's bell, rung by the pilot or the captain, gave various signals. In later years, whistles were used for signalling.

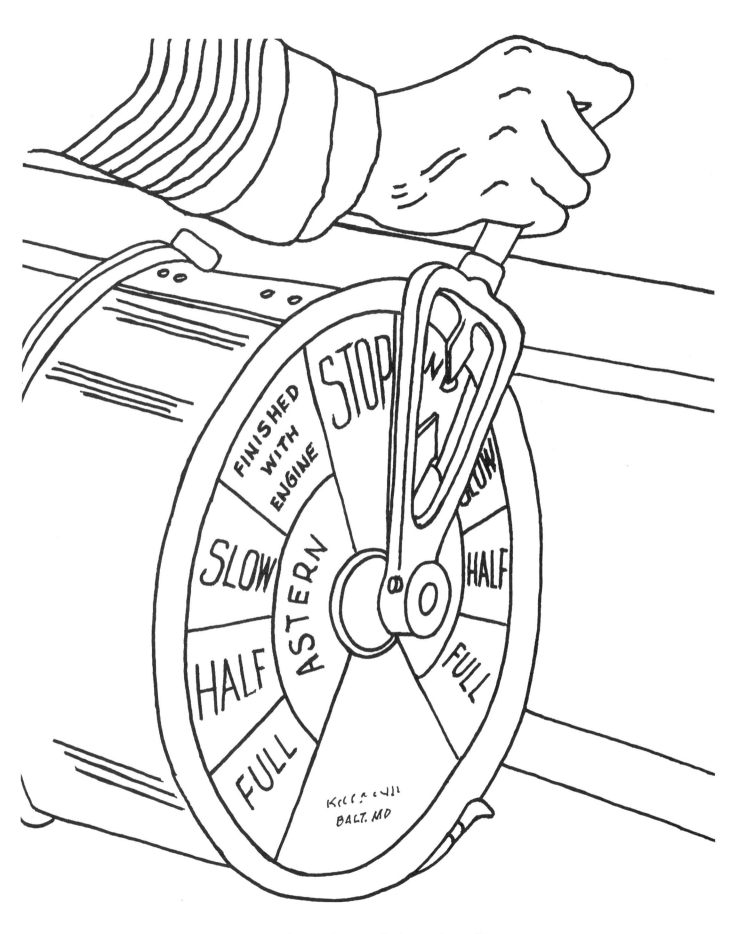

The engine order telegraph is a signalling system that sends orders from the pilot house to the boat's engine room.

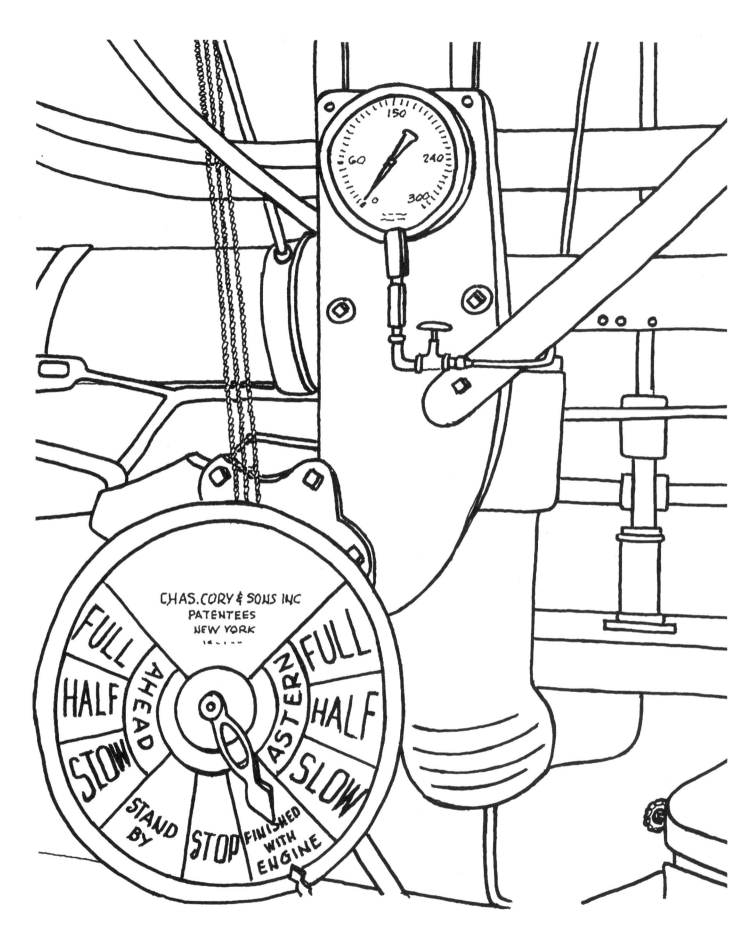

The engine room telegraph receives signals from the pilot house. Other gauges and controls monitor the boat's engines.

The *Delta Queen's* beautiful stairway on the main deck.

Captain Chenegery and banjoist Bud Black entertain passengers aboard the *Delta Queen*.

"Professor" Jay Quimby plays the *Delta Queen* calliope. A calliope is a line of steam whistles that is played like an organ.

The *American Queen,* the grandest steamboat ever!

The *American Queen* steams along the Mississippi.

The *American Queen* has three whistles for signalling.

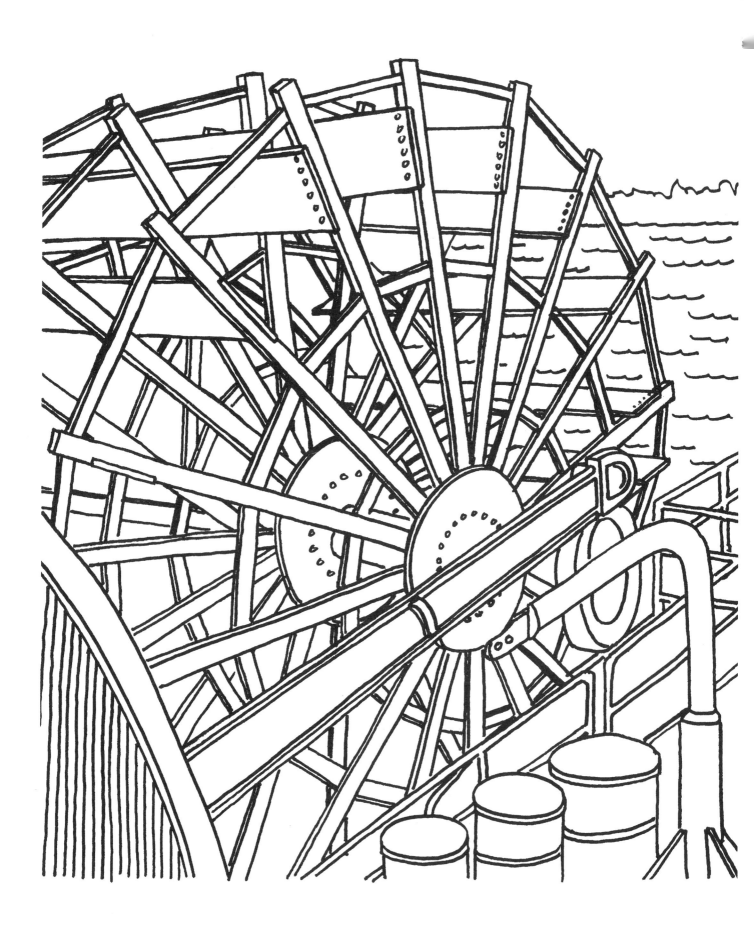

The *American Queen's* paddle wheel, seen from the rear of the main deck.

One of nine boats named *Natchez,* with its stage (gangplank) lowered and ready for bales of cotton to be loaded.

The paddle wheel of the *Mississippi Queen*.

There is a grand collecting of steamboats each year at the "Tall Stacks Festival" in Cincinnati, Ohio.

The famous writer and one-time Mississippi steamboat pilot, Mark Twain.

The Mark Twain boyhood home and museum in Hannibal, Missouri.

Dixieland jazz is featured on the steamboats *Delta Queen, Mississippi Queen,* and *American Queen.*